Neural Nova

Exploring the Cosmic Dust of Human Thought

Lionel Bailey

Neural Nova

*© Copyright 2024 by **Lionel Bailey***

All rights reserved

This document is geared towards providing exact and reliable information with regards to the topic and issue covered. The publication is sold with the idea that the publisher is not required to render accounting, officially permitted, or otherwise, qualified services. If advice is necessary, legal or professional, a practiced individual in the profession should be ordered.

From a Declaration of Principles which was accepted and approved equally by a Committee of the American Bar Association and a Committee of Publishers and Associations.

In no way is it legal to reproduce, duplicate, or transmit any part of this document in either electronic means or in printed format. Recording of this publication is strictly prohibited and any storage of this document is not allowed unless with written permission from the publisher. All rights reserved.

The information provided herein is stated to be truthful and consistent, in that any liability, in terms of inattention or otherwise, by any usage or abuse of any policies, processes, or directions contained within is the solitary and utter responsibility of the recipient reader. Under no circumstances will any legal responsibility or blame be held against the publisher for any reparation, damages, or monetary loss due to the information herein, either directly or indirectly.

Respective authors own all copyrights not held by the publisher.

The information herein is offered for informational purposes solely, and is universal as so. The presentation of the information is without contract or any type of guarantee assurance.

The trademarks that are used are without any consent, and the publication of the trademark is without permission or backing by the trademark owner. All trademarks and brands within this book are for clarifying purposes only and are owned by the owners themselves, not affiliated with this document.

Chapter 1: The Mindscape Galaxy
Neural Star Formation

Thought Constellation Patterns

Deep within the mindscape, thoughts arrange themselves into distinct patterns, much like the stars that form recognizable constellations across the night sky. These thought constellations emerge from the complex interplay of neural activity, creating meaningful formations that guide our understanding and behavior.

Just as ancient civilizations mapped the heavens by connecting stellar points into meaningful shapes, our minds naturally organize thoughts into coherent patterns. These mental constellations serve as navigation tools through the vast expanse of consciousness, helping us make sense of our experiences and memories.

Consider how a musician's thoughts form unique constellations around melody, rhythm, and harmony. When composing, these patterns illuminate connections between different musical elements, creating a symphony of interconnected ideas. Similarly, a mathematician perceives elegant thought patterns linking theorems, proofs, and numerical relationships, forming constellations that reveal hidden mathematical truths.

The brightness of individual thoughts varies within these patterns, much like stars of different magnitudes. Core beliefs and fundamental concepts shine as bright as Sirius or Polaris, while supporting ideas gleam more subtly in the mental firmament. These varying intensities create depth and dimension within each thought constellation, establishing hierarchies of meaning and importance.

Emotional resonance affects the visibility of these patterns. During periods of clarity and insight, thought constellations appear sharp and well-defined, like stars on a crisp winter night. However, when emotions cloud our mental sky, these patterns may become obscured, requiring greater effort to discern their shapes and meanings.

Movement and change characterize these mental formations. Unlike fixed star patterns, thought constellations can shift and evolve as we learn and grow. New experiences add points of light to existing patterns, while others may fade or reposition themselves, creating dynamic mental maps that reflect our developing understanding.

The cultural and personal context influences how these patterns form and connect. Just as different cultures see different images in the same star fields, individuals develop unique thought constellations shaped by their experiences, beliefs, and knowledge. A botanist's mind might organize plant characteristics into one pattern, while an artist sees the same natural elements arranged through aesthetic relationships.

Interaction between thought constellations creates complex networks of meaning. Ideas from different domains form bridges between patterns, like invisible lines connecting distant star clusters. These connections generate innovative insights and creative breakthroughs, as seemingly unrelated concepts align in novel configurations.

The process of learning involves recognizing and mapping new thought constellations. Students gradually discern established patterns within their fields of study while developing personal variations that reflect their unique understanding. These individual interpretations contribute to the collective evolution of knowledge, much like how new star charts build upon ancient astronomical observations.

Memory plays a crucial role in maintaining these patterns. Strong thought constellations become firmly fixed in our mental sky, easily recalled when needed. Others may require regular reinforcement to remain visible, like faint star clusters that demand careful attention to observe.

Problem-solving often involves navigating between different thought constellations, following the connections that link various patterns of understanding. This mental navigation enables us to approach challenges from multiple perspectives, combining insights from different domains to generate innovative solutions.

Dreams and imagination can reveal unexpected thought constellations, temporarily rearranging familiar patterns into new configurations. These nocturnal shifts in our mental sky often provide fresh insights and creative possibilities, expanding our understanding of potential connections between ideas.

The development of expertise in any field involves mastering its characteristic thought constellations. Professionals learn to recognize subtle patterns that remain invisible to novices, developing an intuitive grasp of their domain's mental topography. This deep familiarity enables rapid pattern recognition and informed decision-making.

Through meditation and contemplation, we can observe how these thought patterns form and dissolve. This meta-awareness reveals the dynamic nature of our mental constellations, showing how attention and focus influence their clarity and arrangement. Understanding these processes helps us cultivate more effective thinking patterns and mental habits.

As we continue exploring our inner universe, mapping its thought constellations provides valuable insights into the nature of consciousness and cognition. These patterns serve as both tools for understanding and objects of study themselves, revealing the beautiful complexity of human thought and experience.

Consciousness Nebulae

Vast clouds of awareness swirl through the mindscape, forming intricate patterns that mirror the cosmic nebulae scattered throughout space. These consciousness formations represent the raw material from which thoughts and experiences crystallize, much like stellar nurseries giving birth to new stars.

Within these mental nebulae, consciousness exists in its most primordial state - a mixture of potential thoughts, emerging emotions, and pre-verbal awareness. Like their celestial counterparts, these regions appear differently depending on the observer's perspective and state of mind. Sometimes they manifest as brilliant, illuminated clouds of possibility, while other times they appear as dark, mysterious regions where new ideas gestate.

The dynamics within consciousness nebulae follow patterns similar to their astronomical cousins. Denser regions of awareness attract additional mental energy, creating centers of cognitive gravity that eventually coalesce into fully formed thoughts. These processes occur continuously, with some taking moments while others unfold over years of contemplation and development.

Personal experience shapes the structure and composition of these nebulous regions. Early childhood memories, cultural conditioning, and significant life events create distinctive patterns within the consciousness field. These patterns influence how new experiences are processed and integrated, much like the chemical composition of cosmic nebulae determines the types of stars they can form.

Meditation practitioners often report encountering these consciousness formations during deep states of awareness. They describe witnessing the birth of thoughts from seemingly empty mental space, observing how ideas emerge from the nebulous field of pure potential. This direct observation reveals the delicate interplay between structured thought and formless awareness.

The boundaries between different consciousness nebulae remain fluid and permeable. Ideas and impressions flow between these regions, creating dynamic interactions that spark innovation and insight. Sometimes, these interactions produce sudden bursts of clarity, similar to the brilliant flashes when new stars ignite within cosmic nurseries.

Emotional energy plays a crucial role in shaping these mental formations. Strong feelings can compress and intensify consciousness nebulae, accelerating the development of new thoughts and perspectives. Conversely, periods of emotional calm allow for more gradual and subtle evolution of ideas within these regions.

Creative individuals often learn to navigate these nebulous zones intentionally. Artists, writers, and musicians develop techniques for immersing themselves in these fertile areas of consciousness, drawing inspiration from the raw material of potential. Their work often reflects the patterns and structures they encounter within these mental clouds.

The relationship between individual and collective consciousness creates interesting phenomena within these nebulae. Personal awareness interacts with shared cultural and archetypal patterns, producing hybrid formations that bridge individual and collective experience. These interactions contribute to the evolution of human consciousness and culture.

Sleep and dreams provide special access to consciousness nebulae. During these states, the usual barriers between different regions of awareness become more permeable, allowing for unique combinations and transformations of mental material. Dream experiences often reflect the fluid, nebulous nature of consciousness in its raw form.

Scientific research suggests that consciousness nebulae might correspond to specific patterns of neural activity. When measured, these patterns show similarities to the distribution of matter in cosmic nebulae, hinting at underlying universal principles of organization and development.

The healing potential within these consciousness formations remains largely unexplored. Traditional wisdom traditions suggest that accessing these nebulous regions can facilitate psychological integration and spiritual growth. Modern therapeutic approaches increasingly recognize the value of working with these fluid states of awareness.

Time flows differently within consciousness nebulae. Moments of insight might unfold in apparent slow motion, while extended periods of development can seem to pass in an instant. This temporal flexibility reflects the non-linear nature of consciousness evolution and development.

Understanding these mental formations helps us navigate our inner landscape more effectively. By recognizing the nebulous nature of consciousness, we can work more skillfully with our thoughts and experiences, allowing new insights to emerge naturally from the fertile field of awareness.

The study of consciousness nebulae opens new perspectives on human potential and development. As we learn to work more consciously with these formative regions of mind, we discover fresh opportunities for growth, creativity, and understanding. These discoveries continue to expand our appreciation for the vast and mysterious nature of consciousness itself.

Cognitive Light Years

Distance in the mental universe measures not in miles or kilometers but in degrees of understanding and conceptual leaps. These cognitive light years represent the vast expanses between different states of comprehension, marking the journey thoughts must travel to reach new levels of awareness.

Just as light requires years to traverse the physical cosmos, ideas often need significant time to fully illuminate our understanding. A child grasping basic arithmetic might be separated by several cognitive light years from understanding calculus. Each step of learning bridges these mental distances, gradually closing the gaps between current knowledge and deeper comprehension.

The speed of thought varies dramatically across different mental terrains. Some concepts travel quickly through well-established neural pathways, while others must navigate through dense fields of preconceptions and biases. This variable velocity creates interesting relativistic effects in our mental space-time, where complex ideas might seem to slow down or stretch as they approach the limits of our current understanding.

Personal experience shapes the topology of these mental distances. What appears as a vast cognitive gulf to one person might represent a small step to another, depending on their background, education, and natural inclinations.

Musicians might traverse musical theory effortlessly while finding physics concepts distant and challenging, while physicists might experience the opposite relationship.

Emotional connections can create wormholes through cognitive space, allowing rapid transit between seemingly distant concepts. A powerful personal experience might instantly bridge gaps that would otherwise take years of study to cross. These emotional shortcuts demonstrate the non-linear nature of mental development and learning.

The transmission of knowledge across cognitive light years requires careful translation and interpretation. Teachers act as relay stations, breaking down complex ideas into manageable distances that students can traverse. Each successful transmission reduces the apparent distance between ignorance and understanding, making previously remote concepts accessible.

Memory creates interesting phenomena in this mental spacetime. Recent experiences might feel cognitively closer than they actually are, while deeply understood concepts from long ago maintain their proximity despite temporal distance. This cognitive parallax affects how we navigate our mental universe and access stored information.

Cultural and linguistic differences can add additional light years to cognitive distances. Concepts that seem natural within one cultural framework might require enormous mental journeys to comprehend from another perspective.

These cultural cognitive light years highlight the relative nature of understanding and the importance of context in knowledge transmission.

The development of expertise involves establishing new reference points across cognitive space. As understanding deepens, previously distant concepts move closer, creating new configurations of knowledge that enable further exploration. This continuous recalibration of mental distances characterizes the journey from novice to expert in any field.

Innovation often requires crossing multiple cognitive light years simultaneously, combining distant ideas into novel configurations. Creative breakthroughs happen when seemingly unrelated concepts suddenly appear adjacent in mental space, revealing unexpected connections and possibilities.

The effort required to traverse cognitive distances varies with mental energy and clarity. During peak mental states, we might easily cross distances that seem insurmountable during periods of fatigue or stress. This variability emphasizes the importance of optimal conditions for learning and understanding.

Group collaboration can create cognitive relay networks, allowing ideas to travel faster than any individual mind could manage alone. Scientific communities demonstrate this effect, with researchers building upon each other's work to traverse vast conceptual distances collectively.

The phenomenon of insight represents sudden contractions of cognitive space, where distant understanding suddenly becomes immediately accessible. These moments of clarity collapse apparent light years of mental distance into instant comprehension, though preparing for such breakthroughs often requires extended periods of study and contemplation.

As we age, our cognitive light years shift and evolve. Some mental distances shrink through experience and wisdom, while others might expand as neural plasticity decreases. Understanding these changes helps us adapt our learning strategies and maintain mental flexibility throughout life.

Meditation and contemplative practices can alter our perception of cognitive distances, revealing new pathways through mental space-time. These techniques sometimes uncover shortcuts through seemingly vast expanses of understanding, offering alternative routes to knowledge and insight.

The measurement of cognitive light years provides a useful metaphor for understanding personal growth and development. By recognizing these mental distances, we can better appreciate the journey of learning and the remarkable achievements represented by each advance in understanding. This awareness encourages patience and persistence in our pursuit of knowledge, acknowledging that significant cognitive travels require both time and dedicated effort.

Chapter 2: Cosmic Brain Architecture

Neurological Star Maps

Embedded within the brain's intricate networks lie patterns that mirror the complexity of celestial cartography. These neurological configurations form living maps that guide thought, behavior, and consciousness through the vast territory of human experience. Like ancient astronomers charting the heavens, neuroscientists continue mapping these neural constellations, revealing the beautiful architecture of mind and memory.

The hippocampus serves as our mind's primary cartographer, creating detailed maps of both physical spaces and conceptual landscapes. These neural charts enable navigation through memories and ideas with remarkable precision, storing information in patterns that echo the way celestial maps track stellar movements and relationships.

Each significant experience leaves its mark on these neurological star maps, forming new connections and pathways. Childhood memories, professional expertise, and emotional experiences create distinct patterns of neural activity, like bright clusters of stars marking important regions in our mental sky. These patterns strengthen with repeated use, forming well-traveled routes through our cognitive universe.

The brain's mapping system operates across multiple dimensions, far exceeding the limitations of traditional two-dimensional charts. Memories and knowledge interlink through complex networks of associations, creating multidimensional constellations that connect seemingly unrelated concepts. A simple scent might activate an entire network of memories, much like identifying a single star can help locate entire galaxies of related celestial bodies.

Neuroplasticity ensures these maps remain dynamic and adaptable throughout life. New learning experiences create fresh neural pathways, while unused connections gradually fade, similar to how star maps must be updated to account for stellar movement and evolution. This constant revision allows our internal maps to reflect current needs and circumstances while maintaining critical core knowledge.

The emotional centers of the brain add depth and color to these neural cartographies. The amygdala and limbic system highlight emotionally significant experiences, making them stand out like bright beacons in our mental navigation system. These emotional markers help prioritize information and guide decision-making processes through complex psychological terrain.

Sleep plays a crucial role in maintaining and updating these neurological maps. During deep sleep, the brain consolidates daily experiences into existing neural networks, strengthening important connections and pruning unnecessary ones. This nocturnal recalibration ensures our mental maps remain accurate and efficient, ready to guide us through another day of experience and learning.

Different regions of the brain maintain specialized maps for various aspects of experience. The motor cortex charts physical movements, while language areas map linguistic patterns and meanings. These specialized charts interact and overlap, creating rich, multilayered representations of reality that inform our understanding and behavior.

Trauma and healing can dramatically alter these neural cartographies. Significant emotional events might create disrupted patterns that require careful remapping through therapy and personal growth work. Understanding how these maps can be intentionally revised offers hope for recovery and positive change.

The social brain creates specialized maps of relationships and social dynamics. These charts help navigate complex interpersonal territories, tracking subtle emotional cues and social expectations. Like celestial navigators using stars to cross vast oceans, we use these social maps to guide us through the intricacies of human interaction.

Learning new skills involves creating detailed neural maps of previously uncharted territory. Musicians develop intricate charts of rhythm and harmony, while mathematicians map abstract numerical relationships. These specialized maps become increasingly detailed with practice, allowing for more nuanced and sophisticated navigation within each domain.

Age affects how these neurological star maps function and develop. Children's brains show remarkable flexibility in creating new neural pathways, while older adults might rely more heavily on well-established routes. Understanding these changes helps develop appropriate strategies for learning and adaptation throughout life.

The brain's mapping systems demonstrate remarkable resilience and adaptability. When damage occurs to one region, other areas often compensate by developing alternative pathways, much like navigators finding new routes when familiar stars are obscured by clouds.

Cultural influences shape the development of these neural maps from early childhood. Language, customs, and shared beliefs create common patterns of understanding that influence how individual brains chart experience and meaning. These cultural imprints form the foundation for more personal mappings of reality.

Modern neuroscience continues revealing the incredible sophistication of these biological navigation systems. Advanced imaging techniques show how memories and knowledge are distributed across neural networks, challenging simplified models of brain function and highlighting the dynamic nature of mental mapping.

Understanding these neurological star maps offers practical insights for learning, healing, and personal development. By recognizing how our brains naturally organize and navigate information, we can work more effectively with these innate systems to enhance understanding and promote positive change.

Mental Space-Time Continuum

The fabric of consciousness weaves together space and time in ways that defy conventional physics, creating a unique mental dimension where past, present, and future thoughts coexist and interact. Within this psychological realm, memories of yesterday can feel more immediate than events happening now, while anticipated futures can shape current experiences with gravitational force.

Time flows differently in mental space, expanding or contracting based on emotional intensity and cognitive engagement. A moment of profound insight might stretch into seeming infinity, while hours of routine activity can compress into barely noticeable instances. This elasticity of mental time reflects the subjective nature of consciousness and its relationship to experience.

Spatial relationships in the mind transcend physical limitations. Ideas and memories occupy conceptual locations that can be simultaneously distant and adjacent, depending on the context of thought. A childhood memory might feel spatially closer than yesterday's breakfast, demonstrating how emotional significance warps the geometry of mental space-time.

The architecture of memory creates fascinating distortions in this continuum. Recent experiences often appear larger and more detailed in mental space, while older memories may become compressed yet retain their emotional impact. This compression doesn't necessarily diminish their importance but rather changes their relative position in our psychological landscape.

Attention acts as a powerful force in mental space-time, capable of bending perception and experience. When deeply focused, the mind creates local distortions in subjective time, similar to how massive objects curve space in the physical universe. These attentional fields can slow down perceived time during states of flow or accelerate it during periods of distraction.

Dreams represent unique regions in the mental space-time continuum where normal rules of causality and sequence break down. In these zones, past and future events freely intermingle, and spatial relationships become fluid and symbolic rather than literal. These dream spaces offer glimpses into the full dimensionality of consciousness beyond ordinary waking awareness.

Trauma can create ruptures in mental space-time, forming psychological black holes that distort surrounding experiences. These intense gravitational wells of memory and emotion can trap attention and energy, requiring significant force to escape their pull. Understanding these dynamics helps in developing effective therapeutic approaches for healing psychological wounds.

The social dimension adds another layer of complexity to mental space-time. Shared experiences create overlapping fields of consciousness, where individual perspectives merge and influence each other. Group dynamics can accelerate or slow subjective time, creating collective temporal experiences that differ from personal time perception.

Learning and skill development leave traces through this continuum, forming paths that become easier to traverse with practice. These neural highways allow faster travel between concepts and memories, effectively reducing the psychological distance between different points of understanding. Expertise in any field involves creating efficient routes through mental space-time.

Creativity often emerges from unexpected connections across distant points in this continuum. When ideas from different times and contexts collide, they can generate novel insights and perspectives. These creative syntheses demonstrate the non-linear nature of mental space-time and its potential for innovation.

Meditation practices reveal how consciousness can step outside the normal flow of mental time, accessing states of awareness where past and future lose their usual significance. These experiences suggest that the mind's relationship to time is more flexible than our ordinary perception indicates.

Age influences how we navigate mental space-time, with children experiencing time differently than adults. Young minds often inhabit an expanded present moment, while mature consciousness tends to process experience through denser networks of past references and future projections.

Cultural frameworks significantly shape the structure of mental space-time. Different societies organize temporal experience in unique ways, influencing how their members perceive and interact with past, present, and future. These cultural patterns become embedded in individual consciousness, affecting everything from memory organization to future planning.

The emotional dimension acts as a fourth axis in mental space-time, adding depth and resonance to every point in this continuum. Feelings create affective fields that modify the perceived distance between experiences and alter their temporal relationships. Joy can bring past pleasures into present awareness, while anxiety might pull future concerns closer.

Understanding the nature of mental space-time offers practical applications for personal development and psychological healing. By recognizing how consciousness organizes experience across these dimensions, we can work more effectively with memory, learning, and emotional processing. This awareness allows for more intentional navigation of our inner landscape and better integration of past experiences with present circumstances.

The mental space-time continuum remains one of consciousness's most fascinating aspects, offering endless opportunities for exploration and understanding. As we continue mapping this psychological dimension, we discover new ways to enhance human potential and promote psychological well-being.

Synaptic Solar Systems

Each neuron in the brain operates as a central star, surrounded by intricate networks of dendrites and axons that form complex planetary systems of information exchange. These synaptic solar systems create the fundamental architecture of thought, memory, and consciousness through their dynamic interactions and constant evolution.

The central nucleus of each neuron radiates influence across its domain, much like a sun governing its orbital bodies. Neurotransmitters travel these neural spaces like solar winds, carrying chemical messages that shape the activity and relationships of surrounding cells. This electrochemical dance creates patterns of activation that ripple through entire networks of connected systems.

Dendrites extend like planetary rings around their neuronal stars, collecting input from thousands of neighboring connections. These branching structures create vast reception fields that gather information from multiple sources, integrating diverse signals into coherent patterns of cellular response. The complexity of these dendritic arrangements allows for sophisticated information processing at even the smallest scales.

Synaptic connections form gravitational relationships between neurons, with stronger connections exerting greater influence on their neighbors. These bonds strengthen through repeated activation, creating preferred pathways for information flow that resemble stable orbital patterns in solar systems. Learning and memory formation involve the continuous adjustment of these synaptic gravitational fields.

The birth and death of synaptic connections mirror the dynamic nature of cosmic systems. New connections form like emerging planets, while unused pathways gradually fade like dying stars. This constant renewal and pruning process ensures the brain maintains optimal efficiency while adapting to changing demands and experiences.

Glial cells serve as interstellar medium in these neural cosmos, providing essential support and maintenance for neuronal function. These overlooked components play crucial roles in regulating synaptic transmission, removing cellular debris, and maintaining the delicate balance necessary for healthy brain function.

Neurotransmitter systems create distinct zones of influence within these cellular solar systems. Dopamine, serotonin, and other chemical messengers establish unique domains that modulate neural activity patterns, similar to how different regions of space exhibit varying physical properties and influences on surrounding matter.

The timing of synaptic firing creates complex rhythms that resemble the orbital periods of planetary bodies. These neural oscillations coordinate activity across distant brain regions, enabling coherent thought and consciousness. Different frequency bands serve distinct functions, from memory consolidation to attention and awareness.

Learning involves the formation of new synaptic solar systems and the modification of existing ones. As skills develop, relevant neural networks expand and strengthen, creating more efficient pathways for information processing. This plasticity allows the brain to continuously adapt and evolve in response to experience.

Emotional experiences leave lasting imprints on these cellular configurations. Strong feelings can rapidly reshape synaptic relationships, creating powerful memory traces

psychological skill enables greater freedom in choosing which attractors to engage with and which to avoid.

The interaction between different gravity wells creates complex patterns of psychological movement and development. Understanding these dynamics helps in designing more effective approaches to personal growth, therapy, and educational practices. By recognizing and working with these natural attractors of consciousness, we can better navigate our internal landscape and guide our development in desired directions.

Creative Supernova Events

Moments of extraordinary creative insight explode across the mind like celestial supernovas, releasing enormous bursts of innovative energy that transform the surrounding mental landscape. These rare but powerful events represent the pinnacle of human creative potential, generating waves of novel ideas that ripple through consciousness and culture.

The conditions leading to creative supernovas often build gradually, like the increasing pressure within a massive star. Information, experiences, and ideas accumulate over time, creating cognitive tension that seeks release. When critical mass is reached, the resulting burst of insight can revolutionize entire fields of thought or artistic expression.

Historical records reveal numerous examples of these creative explosions. Mozart's remarkable ability to compose entire symphonies in his mind, Van Gogh's prolific period in Arles, or Einstein's miracle year of 1905 demonstrate how these supernova events can produce extraordinary outputs in compressed timeframes. Such episodes often mark turning points in human cultural and intellectual development.

The neurological basis for creative supernovas involves complex interactions between different brain networks. When the default mode network, associated with daydreaming and internal reflection, synchronizes with executive control systems, the resulting neural harmony can trigger cascades of novel connections and insights. This temporary alignment of usually distinct mental processes enables unprecedented creative synthesis.

Environmental factors play crucial roles in catalyzing these events. Periods of solitude, exposure to novel stimuli, or immersion in challenging problems can create ideal conditions for creative supernovas. Many groundbreaking insights occur during retreats, travels, or times of intense focus on specific challenges.

Sleep often serves as a crucial trigger for creative supernovas. The transitional states between wakefulness and sleep, known as hypnagogic and hypnopompic periods, can release floods of creative associations. Many innovators report receiving their most profound insights during these liminal consciousness states.

Emotional intensity frequently accompanies these creative explosions. Strong feelings can amplify creative energy, leading to periods of exceptional productivity and innovation. The relationship between emotional depth and creative output explains why many breakthrough works emerge during times of personal crisis or extreme joy.

The aftermath of creative supernovas requires careful management. Like their celestial counterparts, these events leave transformed landscapes in their wake. Artists and innovators must learn to harvest and develop the raw material generated during these intense creative periods while maintaining their psychological balance.

Social networks can amplify creative supernovas through collaborative resonance. When multiple creative minds align around shared challenges or inspirations, the resulting synergy can multiply innovative output. Historical creative communities, from the Renaissance workshops to modern research laboratories, demonstrate this multiplicative effect.

Physical movement often catalyzes creative supernovas. Walking, dancing, or other forms of bodily engagement can unlock mental blocks and trigger cascades of insight. This connection between physical and mental movement explains why many creators incorporate regular movement practices into their work routines.

The digital age has introduced new patterns of creative supernova events. Online collaboration platforms and global connectivity enable simultaneous creative explosions across distributed networks of innovators. These distributed supernovas can generate rapid advances in knowledge and artistic expression.

Recovery periods following creative supernovas prove essential for long-term productivity. Like stellar events that temporarily deplete their local environment, intense creative episodes can exhaust mental and emotional resources. Understanding this cycle helps creators maintain sustainable creative practices.

Documentation methods for capturing creative supernova outputs require careful consideration. The speed and intensity of insight during these events often overwhelm normal recording processes. Developing efficient systems for preserving and organizing creative bursts ensures their valuable products aren't lost.

Cultural contexts influence both the frequency and impact of creative supernovas. Societies that support exploration and innovation tend to experience more of these breakthrough events. Understanding these cultural factors helps in creating environments that nurture creative potential.

Age affects the nature of creative supernovas but not their possibility. While younger minds might experience more frequent but less focused creative explosions, mature creators often generate more refined and impactful

breakthroughs. This evolution reflects the interaction
between raw creative energy and accumulated expertise.

Training can increase the likelihood of creative supernovas
through deliberate practice and preparation. Developing
technical skills, building knowledge bases, and maintaining
creative habits creates fertile ground for breakthrough
insights. This preparation enables creators to maximize the
value of these rare events when they occur.

The transformative power of creative supernovas extends
beyond individual achievement to influence entire fields
and societies. These moments of extraordinary insight and
innovation drive human progress, opening new
possibilities for understanding and expression. By
recognizing and nurturing the conditions that support
these events, we enhance our collective creative potential.

Mental Galaxy Clusters

Mental patterns cluster together like vast collections of
galaxies, forming intricate networks of related thoughts,
memories, and behaviors that shape our psychological
universe. These cognitive constellations create the rich
tapestry of human consciousness, each cluster influencing
and interacting with others across the mind's expansive
space.

Professional knowledge forms dense clusters of expertise, where related skills and information naturally gravitate together. A doctor's understanding of human anatomy, disease patterns, and treatment protocols creates an interconnected web of medical knowledge that enables quick diagnosis and decision-making. Similarly, a musician's grasp of theory, technique, and performance practice forms a cohesive cluster that facilitates artistic expression.

Emotional experiences generate powerful clusters that link memories, sensations, and behavioral responses. The joy cluster might connect childhood celebrations, professional achievements, and romantic moments, while the fear cluster binds together past threats, anxious anticipations, and protective responses. These emotional galaxies profoundly influence personality and behavior patterns.

Language creates vast clusters of meaning and association, where words connect to concepts, memories, and cultural references. Bilingual individuals develop parallel linguistic galaxies that sometimes overlap and interact, creating unique patterns of thought and expression. These language clusters shape how we perceive and interact with the world.

Cultural identity forms through the integration of multiple clusters, including shared values, traditions, beliefs, and social practices. These cultural galaxies provide frameworks for understanding and navigating life while influencing everything from food preferences to moral judgments. The interaction between different cultural clusters creates the rich diversity of human society.

Skill development involves the formation of specialized clusters that combine physical coordination, theoretical understanding, and practical experience. Whether learning to play an instrument, master a sport, or practice a craft, these skill galaxies grow more complex and interconnected with practice and dedication.

Memory systems organize into thematic clusters that facilitate recall and understanding. Childhood memories form distinct galaxies from professional experiences, while procedural knowledge clusters separately from declarative facts. These memory organizations enable efficient information retrieval and application.

Social relationships create complex interpersonal clusters, where memories, emotions, and expectations about others form coherent patterns. Family relationships, friendships, and professional networks each develop their own characteristic galaxies of interaction and understanding.

Creative thinking emerges from the novel connections between different mental clusters. When previously separate galaxies of knowledge or experience align in new ways, innovative insights and original ideas emerge. This creative process often involves breaking free from established cluster patterns to forge new connections.

Problem-solving abilities depend on the flexible interaction between various knowledge clusters. Effective solutions often require drawing from multiple domains of expertise and experience, combining elements from different mental galaxies in strategic ways.

Personal values and beliefs form core clusters that influence decision-making and behavior across situations. These fundamental galaxies act as gravitational centers, attracting related thoughts and actions while repelling contradictory influences.

Learning processes create new clusters while modifying existing ones, constantly reshaping the mental landscape. Education involves not just acquiring information but integrating it into meaningful patterns that connect with established knowledge structures.

Trauma can create isolated clusters that remain partially disconnected from normal mental processing. These traumatic galaxies may require therapeutic intervention to gradually reintegrate with healthier thought patterns and emotional responses.

Habits and routines form through the development of automated behavioral clusters that require minimal conscious attention. These efficiency-oriented galaxies enable smooth daily functioning while conserving mental energy for more demanding tasks.

Interest patterns develop as interconnected clusters of curiosity, knowledge, and motivation. These passion galaxies drive personal development and learning while providing sources of meaning and satisfaction in life.

Age influences the organization and interaction of mental clusters, with older individuals often showing more integrated and nuanced pattern recognition across domains. This maturation process reflects the gradual refinement and connection of various knowledge and experience galaxies.

Meditation and mindfulness practices can reveal the structure and movement of these mental clusters, providing insights into their formation and influence. This metacognitive awareness enables more conscious navigation of our psychological universe.

The interaction between different cluster systems creates the unique patterns of individual personality and consciousness. Understanding these complex relationships helps in personal development and professional growth while providing insights into human behavior and potential.

Professional expertise requires cultivating and maintaining specific cluster configurations relevant to one's field. Regular practice and learning ensure these knowledge galaxies remain current and accessible while developing new connections and applications.

The dynamic nature of mental clusters allows for continuous growth and adaptation throughout life. By consciously engaging with and developing these psychological constellations, we can enhance our capabilities and deepen our understanding of ourselves and others.

Chapter 4: Navigating Neural Space

Mental Astronomical Charts

Mapping the mind's celestial patterns reveals intricate constellations of thought, emotion, and behavior that guide our psychological navigation. Like ancient astronomers who traced meaning in the stars, we can chart the recurring formations in our mental sky to better understand and direct our inner experience.

Personal history creates the primary coordinates of our mental chart, with significant life events marking cardinal points that orient all other experiences. Childhood memories form foundational star patterns, while transformative moments appear as bright markers that help navigate present challenges and future aspirations.

Emotional weather systems move across these mental charts in predictable yet dynamic patterns. Joy, sadness, anger, and fear create distinctive atmospheric conditions that influence visibility and navigation. Understanding these emotional climate patterns helps in preparing for and working with different psychological seasons.

Relationship constellations form complex arrangements in our mental space, with family connections, friendships, and professional associations creating distinct but interconnected patterns. These social star maps influence how we relate to others and navigate interpersonal dynamics.

Career trajectories appear as clearly marked paths across the psychological sky, with professional milestones and achievements forming recognizable waypoints. These career constellations guide vocational development and help maintain direction during periods of transition or uncertainty.

Learning patterns trace educational orbits through our mental space, with knowledge clusters forming recognizable formations that facilitate understanding and skill development. These academic constellations grow more detailed and interconnected as expertise deepens in specific areas.

Creative inspiration follows mysterious but mappable routes across the mind's firmament. By tracking these creative pathways, we can better position ourselves to capture innovative insights when they appear on our psychological horizon.

Decision-making processes create decision trees that branch across our mental chart like stellar rivers. Understanding these choice patterns helps in navigating important life crossroads and maintaining consistent direction toward desired goals.

Habit formations appear as well-worn paths across our psychological landscape, some supporting our journey while others leading to less productive territories. Charting these behavioral patterns enables more conscious choices about which routes to maintain or modify.

Cultural influences create background radiation that permeates our entire mental chart, affecting how we interpret and respond to all other patterns. Recognizing these cultural constellations helps in understanding their impact on our psychological navigation.

Memory systems form interconnected star clusters that store and organize past experiences. These mnemonic constellations enable efficient recall while influencing how we perceive and interpret new information.

Value systems appear as fixed points that provide ethical orientation, like moral pole stars guiding behavior and decision-making. These principled constellations help maintain consistent direction despite changing circumstances.

Skill development traces learning curves across our mental chart, with practice and experience creating increasingly refined movement patterns. These expertise constellations enable fluid performance in specific domains.

Emotional intelligence manifests as the ability to read and navigate both personal and interpersonal astronomical patterns. This emotional literacy helps in understanding and working with psychological weather systems effectively.

Time perception creates rhythmic patterns across our mental space, with daily cycles, seasonal changes, and life stages forming recognizable temporal constellations. Understanding these chronological patterns helps in planning and pacing activities.

Personal growth appears as expanding spiral patterns that revisit familiar territories at deeper levels of understanding. These development constellations reveal how we evolve while maintaining connection with our core experiences.

Stress responses create distinctive disturbance patterns in our psychological atmosphere. Recognizing these pressure systems helps in preparing for and managing challenging periods more effectively.

Recovery patterns show how the mind repairs and rejuvenates itself through rest, reflection, and positive experiences. These restoration constellations guide healthy self-care practices and resilience development.

Social support networks appear as interconnected star systems that provide stability and resources during difficult passages. These relationship constellations offer crucial navigation assistance when personal charts become unclear.

Future aspirations form beacon constellations that guide long-term development and goal achievement. These motivational patterns help maintain direction and purpose across life's journey.

By studying and understanding these mental astronomical patterns, we can navigate our psychological space with greater awareness and intention. Like skilled celestial navigators, we learn to read the signs in our mental sky and chart courses that lead toward desired destinations while avoiding known hazards.

Regular observation and mapping of these patterns develops psychological wisdom that enhances decision-making and personal growth. This ongoing process of mental cartography enables more conscious and effective life navigation while deepening our understanding of human experience.

Consciousness Coordinates

Deep within the vast expanse of human awareness lies a sophisticated system of consciousness coordinates that defines our moment-to-moment experience of reality. These coordinates operate across multiple dimensions, creating the unique position from which each person perceives and interacts with the world.

Temporal coordinates anchor our awareness in the flow of time, determining how we experience the present moment while maintaining connections to past memories and future anticipations. This temporal positioning system enables us to navigate through life's chronological landscape, integrating past experiences with current challenges and future aspirations.

Emotional coordinates establish our position within the spectrum of feeling states, from profound joy to deep sorrow, fierce anger to serene calm. These affective markers constantly shift and adjust, responding to both internal and external stimuli while influencing our interpretation of events and relationships.

Sensory coordinates map our physical awareness, creating a detailed grid of bodily sensations, environmental inputs, and spatial relationships. This sensory positioning system grounds consciousness in immediate physical experience while facilitating interaction with the material world.

Cognitive coordinates plot our position within networks of thought, belief, and understanding. These mental markers determine how we process information, make decisions, and construct meaning from our experiences. Their arrangement shapes our intellectual perspective and problem-solving approaches.

Social coordinates locate us within webs of relationships, roles, and cultural contexts. These interpersonal markers influence how we perceive ourselves in relation to others, affecting everything from casual interactions to deep emotional bonds.

Attention coordinates direct the focus of consciousness, determining which aspects of experience receive priority in awareness. This spotlight of attention moves across different coordinate planes, illuminating selected elements while leaving others in shadow.

Memory coordinates connect current experience with stored information, creating continuity in our sense of self across time. These biographical markers enable us to maintain coherent narratives about who we are and how we've developed.

Value coordinates establish our position relative to moral and ethical principles, personal priorities, and life goals. These principled markers guide decision-making and behavior while contributing to our sense of purpose and meaning.

Energy coordinates track our vitality levels, from peak alertness to deep relaxation. These energetic markers influence our capacity for engagement and activity while affecting our overall state of consciousness.

Creative coordinates map our position within spaces of possibility and innovation. These imaginative markers enable us to envision new solutions, express artistic impulses, and generate novel ideas.

Language coordinates position us within systems of symbolic meaning and communication. These linguistic markers shape how we think about and describe our experience while facilitating connection with others.

Identity coordinates establish our sense of self in relation to various roles, characteristics, and capabilities. These personal markers create a stable center from which we engage with life's challenges and opportunities.

Growth coordinates track our development across multiple dimensions of experience. These evolutionary markers reveal patterns of learning, adaptation, and transformation over time.

Spiritual coordinates position consciousness in relation to transcendent experiences and ultimate concerns. These metaphysical markers influence how we understand our place in the larger scheme of existence.

Cultural coordinates locate us within specific historical and social contexts. These contextual markers shape our worldview while influencing our interpretation of events and relationships.

Skill coordinates map our position relative to various capabilities and expertise. These competency markers guide learning and performance while influencing our approach to challenges.

Health coordinates track our physical and psychological well-being. These wellness markers influence overall consciousness while affecting our capacity for engagement and experience.

Purpose coordinates align our awareness with meaningful goals and aspirations. These intentional markers guide long-term development while providing motivation and direction.

Integration coordinates reveal how different aspects of consciousness work together. These synthetic markers show the degree of harmony or conflict between various dimensions of experience.

Understanding these consciousness coordinates enables more effective navigation of inner experience while enhancing our ability to respond skillfully to life's challenges. By recognizing our position across these various dimensions, we can make more conscious choices about where to direct our attention and energy.

Regular monitoring of these coordinates supports personal growth and development while facilitating better self-understanding. Like a skilled navigator using multiple reference points, we can use these consciousness coordinates to chart optimal courses through life's journey.

The dynamic interplay between different coordinate systems creates the rich tapestry of human experience. By developing awareness of these various positioning systems, we enhance our capacity for conscious living while deepening our appreciation for the complexity of human consciousness.

Memory Star Systems

Memory forms vast networks of interconnected star systems, each holding clusters of related experiences, knowledge, and skills that shape our psychological universe. These celestial arrangements of remembered information create the foundation for personal identity and guide future behavior through accumulated wisdom.

Childhood memories form the oldest star systems, their light still influencing our present despite the temporal distance. Early experiences with family, first friendships, and formative events create foundational constellations that shape personality development and emotional patterns. These primal memory clusters often carry the strongest emotional charge, like ancient stars whose gravity affects entire galaxies of later experiences.

Educational memories organize into academic constellations, where subject-specific knowledge forms distinct but interconnected systems. Mathematical concepts orbit alongside scientific principles, while literary understanding and historical knowledge create their own unique patterns of remembered information. These learning clusters grow more sophisticated through continued study and practical application.

Professional memories develop into specialized skill systems, where expertise and experience combine to create efficient patterns of knowledge and capability. Work-related memories form dense clusters of practical wisdom, technical understanding, and problem-solving strategies that enable effective performance in specific domains.

Emotional memories create powerful nebulae of feeling-toned experiences that color current perceptions and influence behavior. Joy, sorrow, fear, and love each generate their own characteristic patterns of remembered sensation and response, forming emotional reference points that guide future reactions.

Relationship memories weave complex networks of shared experiences, creating intimate star systems that define our connections with others. Family bonds, friendships, and romantic relationships each develop unique constellations of shared history, understanding, and expectation.

Procedural memories form automated skill clusters that enable fluid performance without conscious attention. These practical star systems control everything from typing and driving to complex professional techniques, becoming more refined through repeated practice and experience.

Cultural memories connect individual experience to broader social patterns, creating shared constellations of meaning and value. These collective memory systems influence identity formation while providing frameworks for interpreting new experiences.

Traumatic memories can form isolated dark matter within the memory universe, exerting hidden influence on behavior and emotion. These challenging clusters often require special attention and healing to integrate properly with other memory systems.

Achievement memories create bright points of remembered success that illuminate paths toward future accomplishments. These motivational star systems provide confidence and guidance when facing new challenges or setting ambitious goals.

Sensory memories form rich clusters of remembered sights, sounds, smells, tastes, and tactile sensations. These experiential constellations create the textured fabric of recalled experience, bringing past moments vividly into present awareness.

Creative memories combine into innovative patterns that inspire new ideas and solutions. These imaginative star systems enable novel combinations of remembered elements, facilitating original thinking and artistic expression.

Language memories create vast networks of verbal understanding, where words connect to meanings, contexts, and associations. These linguistic constellations enable sophisticated communication while shaping thought patterns and conceptual understanding.

Social memories form intricate webs of interpersonal knowledge, creating sophisticated systems for navigating relationships and group dynamics. These interactive constellations guide behavior in various social contexts while supporting emotional intelligence.

Skills memories cluster into specialized performance systems that enable expert execution in specific domains. These capability constellations grow more refined through practice, creating increasingly sophisticated patterns of remembered expertise.

Future-oriented memories project past experiences forward, creating anticipatory star systems that guide planning and decision-making. These prospective memory clusters help navigate upcoming challenges while supporting goal achievement.

Recovery memories form resilience constellations that provide guidance through difficult times. These coping star systems recall past successes in overcoming challenges, offering hope and practical strategies for current struggles.

Identity memories create core constellations that define personal characteristics and values. These essential star systems maintain consistency in self-understanding while enabling growth and development over time.

Learning memories form educational pathways that facilitate ongoing development. These growth-oriented constellations support the acquisition of new knowledge and skills throughout life.

Integration memories create synthesis between different memory systems, forming bridges between various types of remembered experience. These connecting constellations enable holistic understanding and sophisticated problem-solving approaches. By understanding these memory star systems, we can better appreciate how past experiences shape present functioning while consciously cultivating beneficial memory patterns for future development. Regular engagement with these various memory constellations supports both personal growth and professional excellence.

Chapter 5: The Quantum Mind Universe

Thought Particle Physics

Thoughts behave like fundamental particles in the quantum realm of consciousness, exhibiting both wave-like and particle-like properties as they move through the mind. These cognitive particles interact, combine, and transform in ways that mirror the sophisticated dynamics of physical quantum mechanics.

Individual thoughts possess distinct characteristics of charge, spin, and momentum that influence their behavior and interactions. Positive thoughts carry uplifting energy that tends to attract similar positive particles, while negative thoughts can create dense clusters of pessimistic cognition. The spin of thoughts determines their directional tendency — whether they lead toward action or reflection, expansion or contraction.

Thought particles exist in superposition until observed by conscious attention, containing multiple potential meanings and implications simultaneously. When awareness focuses on a specific thought, this observation collapses its wave function into a more defined state, temporarily fixing its meaning and emotional charge.

Cognitive entanglement occurs when thoughts become interconnected, creating paired systems that influence each other instantaneously regardless of their separation in conscious space. This mental quantum connection explains how triggering one thought immediately activates its entangled partner, forming chains of associated ideas and memories.

The uncertainty principle applies to thoughts as well – the more precisely we try to pin down a thought's exact content, the more its emotional energy becomes uncertain, and vice versa. This fundamental limitation creates a natural fluidity in our thinking processes that resists excessive rigid analysis.

Thought interference patterns emerge when different cognitive waves interact, creating constructive or destructive interference that either amplifies or diminishes certain ideas. These interaction patterns help explain how competing thoughts influence decision-making and emotional states.

Mental quantum tunneling allows thoughts to occasionally bypass normal logical barriers, enabling creative leaps and intuitive insights that seem to emerge spontaneously from the unconscious mind. This phenomenon facilitates innovation and problem-solving through unexpected cognitive connections.

The observer effect manifests in how our attention and expectations influence the behavior and content of thoughts. The very act of monitoring our thinking changes the nature and flow of cognitive particles, creating a dynamic feedback loop between consciousness and thought formation.

Cognitive field effects describe how clusters of related thoughts generate attractive or repulsive forces that influence the movement and organization of other mental particles. These fields shape attention patterns and create preferred pathways for thought flow.

Thought coherence occurs when multiple cognitive particles align their wavelengths and phases, producing powerful states of focused concentration or creative flow. This synchronized mental activity enhances cognitive performance and emotional stability.

The conservation of mental energy governs how thought particles transform and interact while maintaining overall psychic equilibrium. This principle explains why significant cognitive or emotional shifts often require corresponding adjustments in other areas of mental functioning.

Quantum cognitive tunneling enables thoughts to penetrate seemingly impermeable barriers of conditioning or limitation, facilitating breakthrough insights and personal transformation. This mechanism supports learning and psychological growth through unexpected pathways.

Mental wave-particle duality reflects how thoughts can manifest either as discrete units of meaning or as flowing waves of possibility, depending on how they are observed and engaged. This dual nature enables both precise analysis and creative exploration.

Thought resonance occurs when cognitive frequencies align with external stimuli or internal states, creating amplified responses that can significantly influence behavior and perception. Understanding these resonance patterns helps in managing emotional reactions and mental states.

The principle of cognitive superposition allows multiple potential interpretations or solutions to coexist until conscious choice collapses them into specific outcomes. This mental flexibility supports adaptive problem-solving and creative thinking.

Psychological quantum jumps describe sudden shifts in understanding or perspective that occur without intermediate steps, similar to electron orbital changes. These discontinuous transitions facilitate rapid learning and insight generation.

Mental interference patterns emerge from the interaction of different thought waves, creating complex cognitive landscapes that influence decision-making and emotional experience. Understanding these patterns helps in navigating challenging mental states.

The quantum Zeno effect manifests in how frequent observation of thoughts can freeze their natural evolution, explaining how rumination and over-analysis can impede creative flow and emotional processing.

Cognitive entanglement networks form when multiple thoughts become interconnected, creating complex systems of meaning and association that influence broader patterns of thinking and behavior.

By understanding these quantum principles of thought, we can work more effectively with our cognitive processes, fostering states of mental clarity, creativity, and emotional balance. This knowledge enables more sophisticated approaches to personal development and psychological growth while providing practical tools for managing mental states and promoting optimal cognitive functioning.

Neural Wave Functions

Neural waves flow through the brain like intricate symphonies, creating complex patterns that orchestrate every aspect of human consciousness and behavior. These electromagnetic oscillations operate across multiple frequency bands, each serving distinct yet interconnected functions in mental processing and emotional regulation.

Delta waves, the slowest of neural oscillations, dominate deep sleep states and healing processes. These powerful, low-frequency waves sweep across the brain like ocean swells, facilitating physical restoration and memory consolidation. During delta states, the mind processes deep emotional material and integrates daily experiences into long-term memory structures.

Theta waves emerge during meditative states and creative insight, producing a dreamlike consciousness that bridges conscious and unconscious processing. These waves enable access to stored memories and facilitate intuitive understanding, often leading to sudden breakthroughs in problem-solving or artistic expression.

Alpha waves create bridges between external awareness and internal processing, generating states of relaxed alertness optimal for learning and stress reduction. When alpha waves predominate, the mind achieves a balanced state where information flows smoothly between conscious and unconscious levels.

Beta waves activate during focused mental activity and analytical thinking, enabling precise attention and logical processing. These faster oscillations support detailed work, conversation, and decision-making, though excessive beta activity can lead to anxiety and mental fatigue.

Gamma waves, the fastest neural oscillations, coordinate information processing across different brain regions, creating unified perceptual experiences and heightened states of awareness. These waves play crucial roles in learning, memory formation, and peak performance states.

Cross-frequency coupling occurs when different wave patterns synchronize, creating complex harmonics that enable sophisticated mental processes. This neural coordination supports everything from language processing to emotional regulation, allowing seamless integration of various cognitive functions.

Wave coherence measures how effectively different brain regions synchronize their activity, indicating the strength of neural networks and the efficiency of information processing. Higher coherence often correlates with improved cognitive performance and emotional stability.

Phase relationships between neural waves influence how effectively different brain regions communicate and coordinate their activities. Optimal phase alignment supports enhanced memory formation, learning capacity, and emotional regulation.

Amplitude modulation affects the strength of neural signaling, determining how powerfully different brain regions influence each other. This modulation helps direct attention, regulate emotional responses, and control behavioral outputs.

Frequency entrainment occurs when external stimuli guide neural oscillations into specific patterns, enabling targeted enhancement of cognitive functions. This principle underlies various neurofeedback and brain training approaches.

Wave interference patterns emerge when different neural oscillations interact, creating complex effects that can either enhance or inhibit specific mental processes. Understanding these interactions helps optimize cognitive performance and emotional states.

Standing wave patterns form in neural networks, creating stable states of consciousness that support consistent behavior and personality traits. These patterns can be modified through conscious practice and targeted interventions.

Wave propagation through neural tissues follows specific pathways that influence how information flows through the brain. The efficiency of these pathways affects cognitive processing speed and mental flexibility.

Resonant frequencies in neural circuits determine their natural response patterns to different stimuli. Matching these frequencies can enhance learning, memory formation, and emotional processing.

Phase reset mechanisms allow neural networks to quickly shift between different processing states, supporting adaptive responses to changing circumstances. This flexibility enables rapid adjustments in attention and behavior.

Wave amplitude variation creates dynamic patterns that influence emotional intensity and cognitive focus. Managing these variations helps regulate mood and maintain optimal mental states.

Frequency mixing generates combination tones in neural processing, enabling complex information integration and novel insight generation. This mixing supports creative thinking and problem-solving abilities.

Harmonic relationships between neural oscillations create stable processing states that support consistent performance and emotional regulation. Understanding these harmonics helps optimize mental functioning.

The orchestration of these various wave functions creates the symphony of consciousness, enabling the rich complexity of human experience and behavior. By understanding and working with these neural wave patterns, we can enhance cognitive performance, emotional balance, and overall well-being.

Regular practice in modulating these wave patterns through meditation, mindfulness, and other consciousness techniques develops greater control over mental states and emotional responses. This mastery supports personal growth while enabling more effective engagement with life's challenges and opportunities.

Consciousness Field Theory

Consciousness operates as a fundamental field that permeates and influences all aspects of human experience, extending beyond individual minds to create interconnected networks of awareness and meaning. This field manifests through various levels of organization, from personal consciousness to collective awareness, generating complex patterns of thought, emotion, and perception.

The consciousness field exhibits properties similar to electromagnetic fields, with varying intensities, polarities, and frequencies that shape subjective experience. Strong field concentrations create peaks of awareness and clarity, while areas of lower intensity correspond to states of reduced consciousness or unconscious processing.

Individual minds function as nodes within this larger field, each generating unique patterns of thought and emotion that ripple outward to influence the broader consciousness landscape. These personal field patterns interact with those of others, creating complex interference patterns that affect group dynamics and social consciousness.

Field strength varies naturally throughout daily cycles, influencing levels of alertness, creativity, and emotional sensitivity. Peak field intensity often occurs during states of focused attention or profound insight, while reduced field strength characterizes periods of rest or routine activity.

Consciousness field gradients create natural flows of awareness and energy between different states of mind. These gradients guide attention and influence behavior, drawing consciousness toward areas of greater significance or emotional charge while establishing paths of least resistance for thought and action.

Field coherence measures how effectively different aspects of consciousness align and coordinate. Higher coherence supports improved mental clarity, emotional balance, and access to intuitive wisdom, while reduced coherence can manifest as confusion or internal conflict.

The field demonstrates nonlocal properties, enabling instantaneous connections between separated points of consciousness. This nonlocality explains phenomena like intuitive knowing, emotional resonance between individuals, and synchronized group experiences.

Consciousness field boundaries define the limits of individual awareness while simultaneously creating interfaces for interaction with other minds. These boundaries remain semi-permeable, allowing for selective exchange of information and energy while maintaining personal integrity.

Field resonance occurs when different consciousness patterns align in frequency and phase, creating amplified states of awareness and understanding. This resonance facilitates deep communication, shared insight, and collective wisdom emergence.

The field's self-organizing properties enable spontaneous formation of higher-order consciousness structures, supporting the emergence of group minds and cultural consciousness patterns. These organizational dynamics underlie social movements and cultural evolution.

Field memory stores information in distributed patterns throughout the consciousness space, creating an accessible repository of collective wisdom and experience. This memory system supports learning, cultural transmission, and evolutionary development of consciousness.

Consciousness field distortions can arise from trauma, limiting beliefs, or social conditioning, creating areas of restricted flow or blocked awareness. Identifying and resolving these distortions supports psychological healing and personal growth.

Field harmonics emerge from the interaction of different consciousness frequencies, generating complex patterns that influence perception and understanding. These harmonics contribute to aesthetic experience, creative insight, and spiritual awareness.

The field's adaptive properties enable rapid reconfiguration in response to new information or changing circumstances. This flexibility supports learning, problem-solving, and personal transformation while maintaining overall system stability.

Consciousness field amplification occurs through focused attention and intentional practice, enabling enhancement of specific awareness qualities or capabilities. This amplification supports skill development and consciousness evolution.

Field integration processes coordinate different aspects of awareness into coherent wholes, supporting psychological wholeness and mature functioning. This integration enables balanced expression of various personality aspects while maintaining unified identity.

The field's regenerative properties support healing and renewal of consciousness structures, enabling recovery from stress and trauma while facilitating ongoing development and growth. These properties underlie resilience and adaptability.

Consciousness field ecology describes how different awareness patterns interact and influence each other within larger systems. Understanding these ecological relationships supports healthy individual and collective development.

Field resonance between individuals creates channels for deep understanding and connection, facilitating meaningful relationships and effective collaboration. These resonant connections support social cohesion and collective wisdom.

The consciousness field's evolutionary trajectory moves toward increasing complexity and integration, supporting both individual and collective development. This evolution manifests through expanding awareness, deepening wisdom, and growing capacity for love and understanding.

By working consciously with field properties, we can enhance personal development while contributing to collective consciousness evolution. This practical application of field theory supports both individual growth and social transformation through increased awareness and intentional engagement with consciousness dynamics.

Mental Energy States

Mental energy flows through distinct states that profoundly influence our cognitive performance, emotional well-being, and overall life experience. These energetic conditions fluctuate throughout the day, responding to both internal rhythms and external circumstances while shaping our capacity for different types of activity and engagement.

Peak mental energy typically occurs during our natural biological prime time, when cognitive resources align for optimal performance. During these periods, thoughts flow effortlessly, creativity surges, and complex problem-solving becomes more accessible. These states enable deep work and significant breakthroughs in understanding or innovation.

The focused flow state represents a particularly powerful mental energy configuration, characterized by complete immersion in the present moment and seamless engagement with tasks. Time seems to alter its pace, and self-consciousness dissolves into pure activity. This state generates remarkable productivity while actually conserving mental energy through its efficiency.

Contemplative energy states support deep reflection and insight generation. These quieter yet potent conditions allow for meaningful processing of experiences and integration of new understanding. Unlike more active states, contemplative energy moves slowly and deliberately, creating space for wisdom to emerge.

Regenerative mental states occur during periods of conscious rest and recovery. These essential energy configurations allow for neural repair, memory consolidation, and the restoration of cognitive resources. Without adequate time in regenerative states, mental performance deteriorates and emotional balance becomes difficult to maintain.

Social energy states activate during interpersonal interactions, enabling effective communication and emotional connection. These configurations require a unique balance of outward attention and internal processing, drawing on specific mental resources that support relationship building and community engagement.

Creative mental energy manifests as a dynamic, exploratory state that supports innovation and artistic expression. This configuration combines openness to new possibilities with the focused energy needed to bring ideas into form. It often fluctuates between intense activity and receptive stillness.

Analytical energy states enable detailed examination and logical processing. These configurations support systematic thinking and careful evaluation, drawing on mental resources that maintain precision and accuracy while managing complexity.

Emotional energy states color all other mental configurations, influencing how we process information and interact with our environment. These states range from highly charged to subtly nuanced, each affecting cognitive function and behavioral choices in unique ways.

Transitional energy states occur between more defined configurations, serving as bridges that enable smooth shifts in mental functioning. These intermediate conditions help maintain stability during changes in activity or focus, preventing energy loss through abrupt transitions.

Reserve mental energy exists as a backup resource for challenging circumstances or unexpected demands. This energy becomes accessible during emergencies or periods of intense need, though drawing on these reserves requires subsequent recovery time.

Harmonic energy states arise when different aspects of mental functioning align in balanced proportions. These configurations support optimal performance while maintaining sustainable energy use, creating a sense of effortless effectiveness.

Dissonant states emerge when mental energy patterns conflict or compete, creating internal tension and reduced efficiency. Recognizing and resolving these dissonant configurations helps restore effective functioning and well-being.

Boundary energy states maintain healthy separations between different activities and relationships while allowing appropriate connections. These configurations support clear decision-making and effective energy management in complex situations.

Integrative mental states coordinate different aspects of experience into coherent wholes. These configurations enable understanding of complex patterns and relationships while supporting psychological growth and development.

Learning energy states combine receptivity with active processing capacity. These configurations support efficient acquisition and integration of new information while maintaining engagement and motivation.

Restorative states activate healing and repair mechanisms in both mind and body. These essential configurations support recovery from stress and maintain long-term mental health through natural regenerative processes.

By understanding these various mental energy states, we can work more effectively with our natural rhythms and capacities. This knowledge enables better timing of different activities, more efficient use of mental resources, and improved overall performance and well-being.

Regular practice in recognizing and managing mental energy states develops greater control over cognitive and emotional functioning. This mastery supports better decision-making, enhanced creativity, and more effective stress management while promoting sustainable high performance.

The art of mental energy management involves skillful navigation between different states according to circumstances and needs. This dynamic balance supports optimal functioning while maintaining long-term well-being and continued growth.

Cognitive Quantum Entanglement

Cognitive systems exhibit remarkable properties of quantum-like interconnection, creating intricate webs of meaning and influence that transcend classical models of mental processing. These entangled states manifest in human consciousness through various phenomena, from intuitive knowing to synchronized group experiences.

Mental entanglement operates across multiple levels of cognition, linking thoughts, emotions, and memories in ways that defy linear causality. When two or more mental elements become entangled, they form inseparable wholes that respond as unified systems regardless of apparent separation in time or conceptual space.

The phenomenon of cognitive resonance demonstrates how entangled mental states can synchronize across individuals, creating shared understanding and emotional attunement without explicit communication. This explains the often-observed ability of close partners or team members to anticipate each other's thoughts and actions with remarkable accuracy.

Memory systems display entanglement effects through the way associated elements cluster and activate together. A single trigger can instantaneously access entire networks of related memories, emotions, and insights, demonstrating the non-local nature of cognitive information storage and retrieval.

Creative processes harness entangled states when seemingly unrelated ideas suddenly combine to generate novel insights. These spontaneous connections often emerge from deep layers of consciousness where quantum-like processes operate beyond conventional logical structures.

Emotional entanglement creates powerful bonds between experiences, people, and ideas, influencing behavior and decision-making in ways that transcend rational analysis. These emotional linkages can persist across time and distance, maintaining their influence on cognitive processing and behavioral choices.

Learning involves the formation of new entangled states as information becomes integrated into existing knowledge networks. The strength and quality of these entanglements determine how effectively new understanding can be accessed and applied in various contexts.

Problem-solving capabilities often rely on entangled cognitive states that enable simultaneous consideration of multiple possibilities and their interactions. This parallel processing capacity supports complex decision-making and innovative solution generation.

Social cognition demonstrates entanglement through the way individuals in groups develop shared mental models and coordinated behaviors without explicit planning. These collective cognitive fields emerge from the interaction of individual minds in resonant relationships.

Intuitive knowledge often arises from entangled mental states that process information holistically rather than sequentially. This enables rapid understanding of complex situations and relationships without conscious analytical processing.

Language acquisition and use depend on intricate networks of entangled meanings, allowing flexible interpretation and expression across varying contexts. These semantic entanglements support the rich complexity of human communication and cultural transmission.

Skill development creates entangled patterns of perception, thought, and action that enable fluid performance without conscious control. These integrated response patterns become increasingly refined through practice and experience.

Therapeutic change often involves restructuring entangled cognitive-emotional patterns to create healthier response configurations. Understanding these quantum-like properties helps explain why changing single elements often affects entire systems of thought and behavior.

Meditative states can enhance awareness of cognitive entanglement, revealing the interconnected nature of mental processes and enabling more conscious engagement with these quantum-like properties. This expanded awareness supports more effective self-regulation and personal development.

Cultural transmission operates through shared networks of entangled meanings that shape individual and collective behavior. These cultural entanglements influence perception, values, and decision-making at both conscious and unconscious levels.

Leadership effectiveness often depends on the ability to create and maintain productive entangled states within groups, enabling coordinated action and shared vision development. This quantum-like influence operates beyond traditional command and control mechanisms.

Educational processes can be enhanced by working with rather than against natural cognitive entanglement patterns. This approach supports more effective learning by aligning with the mind's inherent tendency to form meaningful connections and associations.

Relationship dynamics demonstrate complex entanglement effects through the way patterns of interaction become established and maintain influence over time. Understanding these quantum-like properties helps explain both the persistence and potential transformation of relationship patterns.

Creative collaboration harnesses collective entangled states to generate innovations that transcend individual capabilities. These shared cognitive fields enable emergence of new ideas and solutions through dynamic interaction of multiple perspectives.

The practical application of cognitive quantum entanglement principles supports enhanced performance across many domains of human activity. By understanding and working with these natural properties of mind, we can develop more effective approaches to learning, problem-solving, and personal development while fostering more productive relationships and collaborative endeavors.

Chapter 6: Mental Astronomical Phenomena

Emotional Solar Flares

Emotional solar flares represent intense bursts of feeling that surge through human consciousness, creating powerful waves of experience that can dramatically affect both individual and collective states of mind. Like their astronomical counterparts, these emotional eruptions carry tremendous energy and can significantly impact the surrounding psychological environment.

The intensity of emotional flares varies considerably, ranging from mild disturbances to overwhelming surges that temporarily disable normal functioning. During peak episodes, individuals may experience heightened sensitivity, altered perception, and dramatic shifts in their capacity to process information or maintain equilibrium.

These emotional phenomena often emerge from deep within the psyche, triggered by both internal and external events that resonate with core psychological patterns. Much like solar activity follows certain cycles, emotional flares tend to occur in patterns that reflect personal and collective rhythms of experience.

When an emotional flare ignites, it typically progresses through distinct phases. The initial buildup may be subtle, with growing tension or unease preceding the main event. As the flare intensifies, emotional energy becomes more concentrated and volatile, often leading to dramatic expressions or profound realizations.

The impact radius of emotional flares extends beyond the individual, affecting others through subtle energetic influences and more obvious behavioral manifestations. Those in close proximity may experience sympathetic responses, creating ripple effects that can influence entire social systems.

Protection from the intense effects of emotional flares requires development of psychological resilience and emotional regulation skills. Just as technological systems need safeguards against solar disturbances, humans benefit from developing internal buffers against overwhelming emotional surges.

Recovery periods following significant emotional flares allow for integration of insights and restoration of balance. These intervals provide essential time for processing experiences and rebuilding emotional resources depleted during intense episodes.

Certain individuals appear more susceptible to emotional flares, displaying heightened sensitivity to both internal and external triggers. This sensitivity can be both a challenge and a gift, offering deeper awareness while requiring careful management of emotional energy.

The transformative potential of emotional flares lies in their ability to break through established patterns and release trapped energy. These breakthrough experiences often lead to significant personal growth and expanded awareness, despite their initially disruptive nature.

Learning to navigate emotional flares skillfully involves developing awareness of early warning signs and understanding personal trigger patterns. This knowledge enables more effective responses and better utilization of the energy released during these events.

The collective dimension of emotional flares becomes evident during periods of social upheaval or shared crisis, when many individuals simultaneously experience intense emotional activation. These synchronized experiences can catalyze significant cultural shifts and evolutionary developments.

Creative expression often channels the energy of emotional flares into productive outcomes, transforming raw emotional power into art, music, writing, or other forms of meaningful communication. This creative transformation helps integrate and give meaning to intense emotional experiences.

The healing potential of emotional flares emerges when their energy helps release old trauma or break through psychological barriers. While potentially overwhelming, these experiences can facilitate deep healing and personal transformation when properly contained and processed.

Environmental factors significantly influence the frequency and intensity of emotional flares, with certain settings either amplifying or dampening their effects. Creating supportive environments helps manage these experiences more effectively while maximizing their potential benefits.

Relationship dynamics often trigger and shape emotional flares, with intimate connections providing both catalysts for activation and containers for processing intense experiences. Understanding these interpersonal dimensions helps navigate emotional flares more successfully within relationships.

Professional settings require particular attention to emotional flare management, as these intense experiences can significantly impact work performance and team dynamics. Developing appropriate strategies for handling emotional flares in professional contexts supports both individual and organizational success.

The wisdom traditions offer valuable perspectives on working with emotional flares, providing frameworks for understanding their significance and techniques for managing their energy. These traditional approaches often emphasize the transformative potential of intense emotional experiences while offering practical guidance for their integration.

Modern research increasingly validates traditional insights about emotional intensity while offering new understanding of the neurological and physiological mechanisms involved. This scientific perspective helps demystify emotional flares while supporting more effective approaches to working with them.

The evolutionary significance of emotional flares lies in their potential to catalyze growth and development, both individually and collectively. By understanding and working consciously with these powerful experiences, we can better harness their transformative potential while minimizing their disruptive effects.

Mastery in handling emotional flares develops through experience and practice, leading to greater emotional intelligence and psychological maturity. This expertise enables more effective navigation of life's challenges while supporting deeper levels of personal and professional development.

Cognitive Aurora Patterns

Cognitive aurora patterns illuminate the mind's landscape with spectacular displays of mental activity, creating dynamic frameworks of thought and perception that shape our understanding of reality. These patterns manifest as intricate waves of consciousness, flowing and shifting like their atmospheric counterparts in the polar skies.

The human mind generates these cognitive displays through complex interactions between different processing systems, creating visible patterns in our conscious experience. Much like the interaction between solar winds and Earth's magnetic field produces auroral displays, our thoughts and perceptions combine to create remarkable mental light shows.

These cognitive patterns vary in intensity and character, sometimes appearing as gentle ripples of insight, other times manifesting as brilliant bursts of understanding that transform our entire mental landscape. Each pattern carries its own signature frequency and quality, influencing how we perceive and interpret our experiences.

During periods of heightened cognitive activity, these patterns become more pronounced, creating spectacular displays of mental illumination. Creative breakthroughs often coincide with particularly vivid cognitive auroras, marking moments when new connections and insights emerge into consciousness.

The rhythmic nature of cognitive aurora patterns reflects underlying neural oscillations that coordinate different aspects of mental processing. These synchronized activities create coherent streams of consciousness that support focused attention, learning, and problem-solving.

Memory formation involves the integration of these patterns into lasting neural networks, with particularly significant experiences creating more vivid and persistent cognitive auroras. These memorable patterns become reference points for future understanding and decision-making.

Emotional states significantly influence the character and intensity of cognitive aurora patterns, adding depth and color to our mental experiences. Joy, sadness, fear, and love each create distinctive patterns that shape how we process and remember events.

Learning to recognize and work with these patterns enhances cognitive performance and emotional well-being. By becoming more aware of our mental displays, we can better navigate complex thinking processes and maintain optimal states for different activities.

Social interactions generate unique cognitive aurora patterns through the resonance between different minds. Group experiences often create synchronized displays that facilitate shared understanding and collective creativity.

The development of expertise in any field involves becoming fluent in reading and generating specific types of cognitive patterns. Masters in various disciplines learn to recognize and work with the subtle variations that characterize their domain of excellence.

Stress and fatigue can distort these patterns, creating interference that disrupts normal cognitive function.

Understanding how to maintain and restore healthy pattern formation supports better mental performance and emotional resilience.

Meditation and contemplative practices often enhance awareness of cognitive aurora patterns, allowing practitioners to observe and influence their mental displays more consciously. This increased awareness supports better self-regulation and deeper insights.

Creative expression often draws directly from these cognitive displays, translating internal patterns into external forms through art, music, writing, and other media. The resulting works can trigger similar patterns in others, facilitating shared understanding and emotional resonance.

Problem-solving abilities depend heavily on the clarity and flexibility of cognitive aurora patterns. When these patterns flow smoothly, solutions often emerge naturally through the dynamic interplay of different mental processes.

The development of wisdom involves learning to read and interpret these patterns with increasing subtlety and depth. This enhanced understanding supports better decision-making and more effective navigation of life's challenges.

Sleep plays a crucial role in maintaining healthy cognitive aurora patterns, allowing for the integration of daily experiences and the restoration of mental clarity. Dream states often display particularly vivid patterns that contribute to psychological processing and creative insight.

Cultural influences shape the interpretation and expression of cognitive aurora patterns, creating shared frameworks for understanding and communicating mental experiences. These cultural patterns facilitate collective learning and social coordination.

The aging process affects how these patterns manifest and evolve, with experience bringing both advantages and challenges in pattern recognition and generation. Understanding these changes helps maintain cognitive flexibility throughout life.

Technology and modern environments influence cognitive aurora patterns in significant ways, sometimes enhancing and sometimes disrupting natural mental rhythms. Developing healthy relationships with technology supports optimal pattern formation and maintenance.

The future of human development may depend largely on our ability to understand and work consciously with these cognitive patterns. As we face increasingly complex challenges, the capacity to generate and interpret clear mental displays becomes ever more crucial for both individual and collective success.

Memory Meteor Showers

Memory meteor showers cascade through consciousness like brilliant streaks of recalled experience, illuminating the mind's sky with fragments of past moments and forgotten insights. These spontaneous bursts of remembrance often arrive in clusters, creating spectacular displays of interconnected recollections that light up our internal landscape.

Just as celestial meteor showers occur when Earth passes through debris fields left by comets, memory showers typically emerge when we encounter powerful triggers that intersect with rich veins of stored experience. A familiar scent, an old song, or a chance encounter can spark these cascading remembrances, each memory igniting the next in rapid succession.

The intensity of these memory showers varies dramatically, from gentle sprinklings of nostalgia to overwhelming storms of recall that temporarily dominate consciousness. During peak events, memories can flood awareness with such force that present reality seems to fade into the background, replaced by vivid reconstructions of past experiences.

Each memory meteor carries its own emotional charge, creating complex patterns of feeling as multiple memories streak through awareness simultaneously. These emotional resonances often combine in unexpected ways, generating new insights and understanding about past experiences and their significance in our lives.

The unpredictable nature of memory showers makes them particularly powerful catalysts for creative inspiration and problem-solving. When diverse memories collide and combine in consciousness, novel connections often emerge, leading to fresh perspectives and innovative solutions to current challenges.

Some memory meteors burn briefly but intensely, leaving lasting impressions despite their momentary nature. Others create longer trails of association, triggering extended sequences of related memories that illuminate previously hidden patterns in our life experience.

The collective dimension of memory showers becomes apparent during shared experiences, when groups encounter common triggers that activate synchronized recall. These shared memory events can strengthen social bonds and contribute to collective understanding through the weaving together of individual recollections.

Certain locations and times seem to generate more frequent memory showers, acting as natural gathering points for accumulated experience. Returning to significant places or marking important anniversaries often triggers these concentrated bursts of remembrance.

The quality of memory meteors varies with age and experience, as newer memories tend to burn brighter while older ones may appear more distant but often carry deeper significance. This natural aging process creates rich layers of meaning as fresh experiences interact with well-worn memory paths.

Processing memory showers effectively requires developing the ability to remain present while allowing recollections to flow naturally. Like watching a celestial meteor shower, the key lies in maintaining a balanced awareness that neither grasps at memories nor pushes them away.

Dream states often feature particularly intense memory showers, as the sleeping mind weaves together fragments of experience into new patterns of meaning. These nocturnal displays can provide valuable insights when properly remembered and interpreted upon waking.

The therapeutic value of memory showers emerges through their ability to bring buried material into consciousness for processing and integration. While sometimes challenging, these spontaneous recalls often facilitate healing by allowing unresolved experiences to be acknowledged and transformed.

Learning to navigate memory showers skillfully enhances both personal growth and professional effectiveness. The ability to work constructively with spontaneous recall supports better decision-making by allowing past experience to inform present choices without overwhelming current awareness.

Memory showers play a crucial role in identity formation and maintenance, as these cascading recollections help weave together the narrative threads of our lives. Each shower contributes to our ongoing story, adding depth and nuance to our understanding of who we are.

The physical environment significantly influences the character and frequency of memory showers, with certain settings more likely to trigger cascading recall. Creating supportive spaces that facilitate healthy memory processing can enhance both personal and professional life.

Cultural factors shape how memory showers are experienced and interpreted, with different societies maintaining varying traditions around remembrance and its significance. Understanding these cultural dimensions helps in working more effectively with memory phenomena across diverse contexts.

The preservation of significant memory showers through journaling, art, or other forms of documentation can create valuable resources for future reference and development. These records often reveal patterns and insights not apparent during the initial experience.

As we age, the character of memory showers evolves, often becoming richer and more complex while potentially losing some immediate clarity. Working consciously with this natural evolution supports healthy cognitive aging and wisdom development.

The integration of memory showers into daily life requires finding balance between honoring past experience and maintaining present engagement. This dynamic equilibrium supports both personal growth and effective functioning in current circumstances.

Understanding and working skillfully with memory showers represents a crucial aspect of psychological development and emotional maturity. By learning to navigate these spontaneous cascades of remembrance, we can better harvest their insights while maintaining stable ground in present awareness.

www.ingramcontent.com/pod-product-compliance
Lightning Source LLC
Chambersburg PA
CBHW061330120726
48001CB00002B/777